Reflections on the Human Condition

Books by Eric Hoffer

THE TRUE BELIEVER
THE PASSIONATE STATE OF MIND
THE ORDEAL OF CHANGE
THE TEMPER OF OUR TIME
WORKING AND THINKING ON THE WATERFRONT
FIRST THINGS, LAST THINGS
REFLECTIONS ON THE HUMAN CONDITION

Reflections on the Human Condition

ERIC HOFFER

PERENNIAL LIBRARY
Harper & Row, Publishers
New York, Evanston, San Francisco, London

Portions appeared in the September 24, 1972, issue of *Family Weekly*.

A hardcover edition is available from Harper & Row, Publishers.

REFLECTIONS ON THE HUMAN CONDITION. Copyright © 1973 by Eric Hoffer. All rights reserved. Printed in the United States of America. No part of this book may be used or reproduced in any manner whatsoever without written permission except in the case of brief quotations embodied in critical articles and reviews. For information address Harper & Row, Publishers, Inc., 10 East 53rd Street, New York, N.Y. 10022. Published simultaneously in Canada by Fitzhenry & Whiteside Limited, Toronto.

First PERENNIAL LIBRARY edition published 1974.

STANDARD BOOK NUMBER: 06-080309-6

To Lili

Contents

I	Between the Dragon and the Devil	1
II	Troublemakers	27
III	Creators	47
IV	Prognosticators	65
V	The Individual	77

I

*Between the Dragon
and the Devil*

1

Nature attains perfection, but man never does. There is a perfect ant, a perfect bee, but man is perpetually unfinished. He is both an unfinished animal and an unfinished man. It is this incurable unfinishedness which sets man apart from other living things. For, in the attempt to finish himself, man becomes a creator. Moreover, the incurable unfinishedness keeps man perpetually immature, perpetually capable of learning and growing.

2

There is something unhuman about perfection. The performance of the expert strikes us as instinctual or mechanical. It is a paradox that, although the striving to master a skill is supremely human, the total mastery of a skill approaches the nonhuman. They who would make man perfect end up by dehumanizing him.

3

The source of man's creativeness is in his deficiencies; he creates to compensate himself for what he lacks. He became Homo faber—a maker of weapons and tools—to compensate for his lack of specialized organs. He became Homo ludens—a player, tinkerer, and artist—to compensate for his lack of inborn skills. He became a speaking animal to compensate for his lack of the telepathic faculty by which animals communicate with each other. He became a thinker to compensate for the ineffectualness of his instincts.

4

Man was nature's mistake—she neglected to finish him—and she has never ceased paying for her mistake. For it was in the process of finishing himself that man got out from underneath nature's inexorable laws, and became her most formidable adversary.

5

Due to the imperfection of man's instincts there is a pause of faltering and groping between his perception and action. This pause is a seedbed of the apprehensions, the insights, the images, and the concepts which are the warp and woof of the creative process. A shrinking of the pause results in some degree of dehumanization. This is as true of highly trained specialists and dogmatic True Believers as of the mentally deficient.

Both iron discipline and blind faith strive to eliminate the pause of hesitation before action, while the discipline that humanizes and civilizes aims at widening the interval between impulse and execution.

Art humanizes because the artist must grope and feel his way, and he never ceases to learn.

6

You dehumanize a man as much by returning him to nature—by making him one with rocks, vegetation, and animals—as by turning him into a machine. Both the natural and the mechanical are the opposite of that which is uniquely human. Nature is a self-made machine, more perfectly automated than any automated machine. To create something in the image of nature is to create a machine, and it was by learning the inner working of nature that man became a builder of machines. It is also obvious that when man domesticated animals and plants he acquired self-made machines for the production of food, power, and beauty.

7

Animals often strike us as passionate machines.

8

When you automate an industry you modernize it; when you automate a life you primitivize it.

Man is godlike when he makes nature pliable and responsive, but becomes an anti-God when he turns human beings into a plastic mass that he can knead at will. For God turned clay into man, while the anti-God turns man into malleable clay.

9

The mindlessness of nature frightens us, particularly when we are made aware of the ingenuity and precision by which it achieves its ends. That chance should accomplish, even over immeasurably long periods and with immeasurable waste, what only the subtlest mind could devise seems to us beyond belief. We find it easier to believe in God than in the purposeful working of blind chance.

10

Nature is always rational. Every answer you pry from nature is severely logical. When the wind turns into a tornado it does so not by irrational madness but by a mathematically precise process. It seems paradoxical that that which has no mind should be unfailingly rational. The slang expression "mental" for the insane points to the fact that irrationality has its source in the mind; that it is a reaction against the intellect.

11

Mass movements use irrationality to shut out the intellect, to turn people into predictable, mindless machines. Both Stalin and Hitler used blind faith as a device for mechanizing souls.

12

We hear a lot about the dehumanizing effects of the machine. Actually, the large-scale dehumanization of the Stalin-Hitler era was the work of ideological machines. In Russia the doctrinaire appliances work better than the mechanical.

13

The savior who wants to turn men into angels is as much a hater of human nature as the totalitarian despot who wants to turn them into puppets.

There are similarities between absolute power and absolute faith: a demand for absolute obedience; a readiness to attempt the impossible; a bias for simple solutions—to cut the knot rather than unravel it; the viewing of compromise as surrender; the tendency to manipulate people and "experiment with blood." Both absolute power and absolute faith are instruments of dehumanization. Hence absolute faith corrupts as absolutely as absolute power.

14

The fanatic deals with men the way the scientist deals with matter. There is a startling similarity between Bacon's prescription for mastering nature—"Nature, to be commanded must be obeyed"—and Loyola's formula for manipulating men—"Follow the other man's course to your own goal."

15

Man is utterly fantastic when seen as an animal, a god, a machine, or a physiochemical complex. Nothing makes man so fantastically unfamiliar as when he is likened to something familiar. To forget that man is a fantastic creature is to ignore his most crucial trait.

Take man's most fantastic invention—God. Man invents God in the image of his longings, in the image of what he wants to be, then proceeds to imitate that image, vie with it, and strive to overcome it. In the Occident, the revolt against God released undreamed-of energies. It was the rejection and usurpation of a once ardently worshiped God that had fateful effects on society and the individual in the modern Occident.

16

When we renounce a faith we do not cast it off but swallow it: we substitute the self for the abandoned holy cause, and the result is not lethargy but an intensification of the individual's drive.

17

Belief passes, but to have believed never passes.

18

Man made God in his own image. In whose image did he make the devil? The devil with hoofs, tail, and horns is obviously a beast masquerading as a man. Does he, then, personify nature? Is there a confrontation—God and man on one side, the devil and nature on the other?

It is significant that where men live in awe of nature and see it as inexorable and inscrutable fate, nature is personified not in a devil but in a dragon. The dragon is a composite of the fearsome strengths and uncanny faculties of the animal world. Any piecing-together of parts of various animals will result in something like a dragon. Vasari tells how the young Leonardo da Vinci, wanting to paint something that would frighten everybody, brought to his room every sort of living creature he could lay his hands on and set out to paint a composite animal. "He produced an animal so horrible and fearful that it seemed to poison the air with its fiery breath. This he represented coming out of some dark broken rocks, with venom issuing from its jaws, fire from its eyes, and smoke from its nostrils. A monstrous and horrible thing indeed." In the course of time the dragon came to embody the menace and the mystery of the whole nonhuman cosmos. "The dragon," says Kakuzo Okakuro, "unfolds himself in the storm clouds, he washes his mane in the blackness of the seething whirlpool. His claws are in the forks of lightning, and his scales begin to glisten in the bark of rainswept pine trees. His voice is heard in the hurricane." Since societies awed by nature tend to equate power with nature, they will invest omnipotent individuals—emperors, despots, warriors, sorcerers, etc.—with the attributes of the dragon. Thus, unlike the devil, the dragon is a man masquerading as a beast.

The dragon is infinitely more ancient than the devil. The earliest representation of the dragon is the painting of the sorcerer in the cave of Trois Frères. This Late Paleolithic painting presents a sorcerer decked out as a composite animal with the horns of a reindeer, the ears of a wolf, the eyes of an owl, the paws of a bear, and the tail of a horse.

The devil is coetaneous with Jehovah, the God who is not nature but its creator. It was the feat of the ancient Hebrews that though without an advanced technology they lost their awe of nature, and saw it as man's task to "subdue the earth." And once man, backed by Jehovah or a potent technology, assumes a cocky attitude toward nature, the devil comes upon the scene and takes the place of the dragon. The devil personifies not the nature that is around us but the nature that is within us—the infinitely ferocious and cunning prehuman creature that is still within us, sealed in the subconscious cellars of the psyche.

Outside the Occident, where nature has the upper hand, the dragon is still supreme, but the Occident proper is the domain of the devil.

It is of interest that at his first appearance in the Garden of Eden, before clothes were invented, the devil came undisguised, and contrived the fall of man from a paradisiacal existence. Nowadays the devil is decked out in the latest fashion, and quotes the latest scriptures.

We of the present are vividly aware that the slaying of the dragon is the opening act in a protracted, desperate contest with the devil. The triumphs of the scientist and the technologist are setting the stage for the psychiatrist and policeman. We also know that we can cope with the devil only by using the tension between that which is most human and nonhuman in us to stretch souls in a creative effort.

19

Deep probings into man's nature invariably come up with scandalous evidence of his innate vileness.

The historian Friedrich Meinecke was so disconcerted by the dark and impure origins of great cultural values that it seemed to him as though "God needed the devil to realize himself." Yet, considering man's origins, the startling thing is not the evil that is at the root of cherished values, but the alchemy of the soul which transmutes unflagging malice and savagery into charity, love, and visions that reach unto heaven. For the prehuman creature from which man evolved was unlike any living thing in its malicious viciousness toward its own kind. Had it not been for the appearance of a mutation of sociableness, by means of babbling, laughter, and the dance, the species would have perished. Humanization was not a leap upward but a groping toward survival. Original sin has its roots in man's origins: we are descended from a devil. And since it still holds true that man is mankind's deadliest enemy, the survival of the species still depends on further humanization.

Until we become wholly human we are all to some extent devils—beasts masquerading as men.

20

Laughter to begin with was probably glee at the misfortunes of others. The baring of the teeth in laughter hints at its savage ancestry. Animals have no malice, hence also no laughter. They never savor the sudden glory of *Schadenfreude*. It was its infectious quality which made of laughter a medium of mutuality.

Beasts are not beastly. The evil of dehumanization is not that it turns us into animals but turns us into the malignant prehuman monstrosity from which man evolved.

21

The capacity for transcending the senses—for telepathic transmission and for sensing the unseen—is an animal characteristic. It is doubtful whether the mutation of passionate sociableness, which started the process of humanization, would have been possible without a blunting of the capacity to sense evil intentions. It is still true that a misunderstanding takes place not when people fail to understand each other, but when they sense what is going on in each other's mind and do not like it. Pascal feared that if men knew what each thought of the other there would be no friends in the world.

22

One wonders whether it was the alchemy of the soul—the fact that in man's soul good and evil, beauty and ugliness, truth and error continually pass into each other—that gave rise to the idea of an alchemy in nature. When the alchemists tried to transmute one metal into another, they were trying to deal with nature as if it were human nature.

23

Has the dread of a nuclear holocaust brought the dragon to the Occident? A generation made aware of a nuclear threat from birth seems to have a superstitious awe of nature. The ecological fervor of the young is the manifestation of an urge to propitiate nature. So, too, the revival of astrological superstitions and the receptivity to Asian cults betoken a changed attitude toward nature.

24

How does a vivid awareness of the evil that is in us affect a person's view of the world?

Most of the people who delved deeply into man's nature were not overly disconcerted by the discovery that ill will and hatred are all-pervading ingredients in the compounds and combinations of our inner life. Montaigne, Bacon, La Rochefoucauld, Hume, Renan, and others derived an exquisite delight from tracing and identifying the questionable motives which shape human behavior. Pascal saw it as evidence of divine grace that an impulse of charity is distilled from the evil brew that simmers in men's souls. In John Calvin the combination of intense self-awareness with a fervent belief in God had outlandish results.

Calvin knew beyond doubt that there was no such thing as a genuinely good deed. "No work of pious men ever existed which, if it were examined before the strict judgment of God, did not prove damnable." The jealous, malicious, all-hating "I" thrives on the misfortunes of others, and this "I" is the bedrock on which virtue and piety rest. Calvin had, therefore, to assert categorically that we cannot earn salvation by good deeds. From this it was but a step to the preposterous doctrine of predestination.

25

It is not only more sensible but more humane to base social practice on the assumption that all motives are questionable and that in the long run social improvement is attained more readily by a concern with the quality of results than with the purity of motives. The establishment of a desirable pattern of habits is more vital than the implanting of right beliefs and motives. A concern with right and wrong thinking is the manifestation of a primitive, superstitious mentality.

26

Good and evil grow up together and are bound in an equilibrium that cannot be sundered. The most we can do is try to tilt the equilibrium toward the good.

27

That man, a deficient, lesser animal, became more than an animal was due to his singular endowment for turning handicaps into advantages. Man's tools and weapons more than made up for his lack of specialized organs, and his capacity for learning did more for him than inborn skills and organic adaptations. It still holds true that man is most uniquely human when he turns obstacles into opportunities.

The most fateful consequence of man's incurable unfinishedness is his chronic immaturity, his inability to grow up, his perpetual youthfulness. By all odds, earliest man, so naked to the elements and to deadly enemies, should have existed in a state of constant shock. We find him instead the only lighthearted being in a deadly serious universe. All around him were living creatures superbly equipped, and driven by grim purposefulness. He alone, with childish carelessness, tinkered and played, and exerted himself more in the pursuit of superfluities than of necessities. Yet the tinkering and playing, and the fascination with the nonessential, were a chief source of the inventiveness which enabled man to prevail over better-equipped and more-purposeful animals.

Most of the time when we try to trace the ancestry of a device or a practice which played a crucial role in the ascent of man we reach the realm of the nonutilitarian. Many weapons and tools were to begin as playthings. The bow was a musical instrument before it became a weapon, and the wheel was a plaything before it was used as a tool. The Aztecs had no wheels, but many of their toys had rollers for feet. Clay figurines antedated clay pots, and ornaments came before clothes. The first domesticated animals were pets, and it has been suggested that grain was first cultivated not to raise food but to make beer. In recent centuries many machines made their first appearance as mechanical toys.

The Paleolithic hunters who painted the unsurpassed animal murals on the ceiling of the cave at Altamira had only rudimentary tools. Art is older than production for use, and play older than work. Man was shaped less by what he had to do than by what he did in playful moments. It is the child in man that is the source of his uniqueness and creativeness, and the playground is the optimal milieu for the unfolding of his capacities and talents.

28

Man is a luxury-loving animal. Take away play, fancies, and luxuries, and you will turn man into a dull, sluggish creature, barely energetic enough to obtain a bare subsistence. A society becomes stagnant when its people are too rational or too serious to be tempted by baubles.

29

To be aware how fruitful the playful mood can be is to be immune to the propaganda of the alienated, which extols resentment as a fuel of achievement.

30

It is a juvenile notion that a society needs a lofty purpose and a shining vision to achieve much. Both in the marketplace and on the battlefield men who set their hearts on toys have often displayed unequaled initiative and drive. And one must be ignorant of the creative process to look for a close correspondence between motive and achievement in the world of thought and imagination.

31

The romantic view of history which sees grandiose conceptions at the root of great deeds stems from an unrealized passion for impressive action just as the romantic view of love thrives on sexual frustration.

32

The central task of education is to implant a will and facility for learning; it should produce not learned but learning people. The truly human society is a learning society, where grandparents, parents, and children are students together.

In a time of drastic change it is the learners who inherit the future. The learned usually find themselves equipped to live in a world that no longer exists.

33

It is the malady of our age that the young are so busy teaching us that they have no time left to learn.

34

Since man has to finish and "make" himself, there are unavoidably greater differences between individual men than between individual animals. The *Gleichschaltung* of individuals always results in some degree of dehumanization. This is true even when individuality is sacrificed for a declared common good.

One must also expect chance to play a greater role in the lives of men than in the lives of animals. It is true that where sheer survival is concerned accidents are less decisive in man than in other forms of life. Much of the time society shields a person against death by accident. But in the unfolding of the individual's life, chance is everything. In a vigorous society chance and example have full play, and in such a society the talented are likely to be lucky.

35

Man is the most alive of living things. In him the traits which distinguish the animate from the inanimate become most pronounced. This is particularly true of his creativeness, which is essentially life giving. It introduces order into the randomness of nature, builds associations which qualitatively transcend the constituent parts, and is actuated not only by the present environment but by memories and goals.

36

In the alchemy of man's soul almost all noble attributes—courage, honor, love, hope, faith, duty, loyalty, etc.—can be transmuted into ruthlessness. Compassion alone stands apart from the continuous traffic between good and evil proceeding within us. Compassion is the antitoxin of the soul: where there is compassion even the most poisonous impulses remain relatively harmless.

Nature has no compassion. It is, in the words of William Blake, "a creation that groans, living on death; where fish and bird and beast and tree and metal and stone live by devouring." Nature accepts no excuses and the only punishment it knows is death.

37

Man started out as a "weak thing of the world" and evolved "to confound the things that are mighty." And within the human species, too, the weak often develop aptitudes and devices which enable them not only to survive but to prevail over the strong. Indeed, the formidableness of the human species stems from the survival of its weak. Were it not for the compassion that moves us to care for the sick, the crippled, and the old there probably would have been neither culture nor civilization. The crippled warrior who had to stay behind while the manhood of the tribe went out to war was the first storyteller, teacher, and artisan. The old and the sick had a hand in the development of the arts of healing and of cooking. One thinks of the venerable sage, the unhinged medicine man, the epileptic prophet, the blind bard, and the witty hunchback and dwarf.

38

One wonders whether boundless, all-embracing compassion for our fellow creatures might not enable us to tackle the apparently insoluble problems of a time of drastic, rapid change. Up to now, whenever a society turned a new leaf it had the devil at its elbow.

II

Troublemakers

39

Everywhere we look at present we see something new trying to be born. A pregnant, swollen world is writhing in labor, and everywhere untrained quacks are officiating as obstetricians. These quacks say that the only way the new can be born is by a Caesarean operation. They lust to rip the belly of the world open.

40

It is maintained that a society is free only when dissenting minorities have room to throw their weight around. As a matter of fact, a dissenting minority feels free only when it can impose its will on the majority: what it abominates most is the dissent of the majority.

41

There is a spoiled-brat quality about the self-consciously alienated. Life must have a meaning, history must have a goal, and everything must be in apple-pie order if they are to cease being alienated. Actually, there is no alienation that a little power will not cure.

42

The untalented are more at ease in a society that gives them valid alibis for not achieving than in one where opportunities are abundant. In an affluent society, the alienated who clamor for power are largely untalented people who cannot make use of the unprecedented opportunities for self-realization, and cannot escape the confrontation with an ineffectual self.

In America, the post-Sputnik education explosion has produced a horde of would-be uncommon people who want to live uncommon lives and do uncommon deeds. Hence the present appetite for historymaking on campuses and in intellectual cliques.

43

We all have private ails. The troublemakers are they who need public cures for their private ails.

44

Commitment becomes hysterical when those who have nothing to give advocate generosity, and those who have nothing to give up preach renunciation.

45

We are told that the paradise of the masses in America is a "pig heaven." Granted; but the Europe the masses left behind was a pigsty.

46

It is cheering to see that the rats are still around—the ship is not sinking.

47

Right now it seems that they who have a truth to reveal also have a lie to hide.

48

Dissenting intellectuals have had it best in middle-class societies—societies they abominate and try hard to destroy. They had it worse in societies dominated by intellectual hierarchies—the hierarchies of secular and sacerdotal churches.

49

Though dissenters seem to question everything in sight, they are actually bundles of dusty answers and never conceived a new question. What offends us most in the literature of dissent is the lack of hesitation and wonder.

50

Nonconformists travel as a rule in bunches. You rarely find a nonconformist who goes it alone. And woe to him inside a nonconformist clique who does not conform with nonconformity.

51

He who proselytizes in the cause of unbelief is basically a man in need of belief.

52

It almost seems that nobody can hate America as much as native Americans. America needs new immigrants to love and cherish it.

53

The history of this country was made largely by people who wanted to be left alone. Those who could not thrive when left to themselves never felt at ease in America. This is true not only of the pampered rich, but of the intellectuals, the chronically poor, and to some extent of the Negro.

54

Anyone aware of the imperfections inherent in human affairs is hardly capable of total commitment. Part of him will inevitably remain uncommitted. It is this perch of uncommitment which makes an act of self-sacrifice sublimely human, and distinguishes the man of faith from the fanatic.

55

We take for granted the need to escape the self. Yet the self can also be a refuge. In totalitarian countries the great hunger is for private life. Absorption in the minutiae of an individual existence is the only refuge from the apocalyptic madhouse staged by maniacal saviors of humanity.

56

One of the chief objectives of freedom is to make it possible for a person to feel himself a human being first. Any social order in which people see themselves primarily as workingmen, businessmen, intellectuals, members of a church, nation, race, or party is deficient in genuine freedom.

57

To some, freedom means the opportunity to do what they want to do; to most it means not to do what they do not want to do. It is perhaps true that those who can grow will feel free under any condition.

58

As crucial as what happens to an individual at birth and in earliest childhood is what happens to him at rebirth, during his passage to adulthood. Moreover, the manner in which a generation passes from childhood to manhood can have fateful effects on society as a whole.

If a society is to preserve its stability and a degree of continuity, it must know how to keep its adolescents from imposing their tastes, attitudes, values, and fantasies on everyday life. At present, most nations are threatened more by their juveniles within than by enemies without.

In America the post-Sputnik decade saw an unprecedented widening of the age range of adolescence. Television has given ten-year-olds the style of life of juveniles, while the education explosion has kept people in their late twenties cooped up as students on campuses. The spectacular expansion of the adolescent group combined with a failure of nerve in the adult population has profoundly changed the nature of our society. Anything written about America as recently as ten years ago has become out of date.

Perhaps a modern society can remain stable only by eliminating adolescence, by giving its young, from the age of ten, the skills, responsibilities, and rewards of grownups, and opportunities for action in all spheres of life. Adolescence should be a time of useful action, while book learning and scholarship should be a preoccupation of adults. It would be also particularly fitting were the training and coaching of the young done by retired skilled craftsmen, technicians, industrialists, scientists, and politicians.

59

Is there a self-aware person who does not squirm when he remembers what he was like at twenty?

A compilation of what outstanding people said or wrote at the age of twenty would make a collection of asinine pronouncements. "When I remember what I was like between seventeen and twenty-seven," said Henri de Montherlant, "I wish I could spit at my former self."

60

One wonders whether a generation that demands instant satisfaction of all its needs and instant solution of the world's problems will produce anything of lasting value. Such a generation, even when equipped with the most modern technology, will be essentially primitive—it will stand in awe of nature, and submit to the tutelage of medicine men.

61

In human affairs every solution serves only to sharpen the problem, to show us more clearly what we are up against. There are no final solutions.

62

Action is released by emotion, and emotion is stirred by words. What, then, is the role of thought in the release of action? For all we know its role is as an instrument in the production of potent words.

63

The link between ideas and action is rarely direct. There is almost always an intermediary step in which the idea is overcome. De Tocqueville points out that it is at times when passions start to govern human affairs that ideas are most obviously translated into political action. The translation of ideas into action is usually in the hands of people least likely to follow rational motives. Hence it is that action is often the nemesis of ideas, and sometimes of the men who formulated them.

One of the marks of a truly vigorous society is the ability to dispense with passion as a midwife of action—the ability to pass directly from thought to action.

64

The nineteenth century planted the words which the twentieth ripened into the atrocities of Stalin and Hitler. There is hardly an atrocity committed in the twentieth century that was not foreshadowed or even advocated by some noble man of words in the nineteenth.

65

There are times when words and ideas are deemed dangerous, and times when people may profess the most incendiary ideas and shout them on housetops without anyone being alarmed.

Obviously, ideas are deemed dangerous when things are in flux. Thus in the formative days of Christianity and during the crisis which gave birth to the modern Occident ideas and words were seen as a sort of explosive. The same is true of our time, when nations and social orders are in a state of transition.

Still, it is the close link with deeds that renders words dangerous, and such a link does not come automatically into existence by a fluidity of conditions. In addition to the criticalness of the times there must also be present a relatively large number of a certain type of "men of words" who hover on the borderline between words and action. This category of intellectuals is made up of people who are basically men of action—potential managers, administrators, organizers—who, owing to circumstances, find themselves encased in the career of "men of words." In a country like France, for instance, where the social landscape is tilted toward letters and the arts, many potential business tycoons are washed into the career of intellectuals. So, too, in this country, since Sputnik, the prestige and rewards of intellectual pursuits have risen so sharply that many individuals with superb talents for wheeling and dealing and for building industrial empires are now throwing their weight around in the universities. These people want to act, command, and make history, but being cast in the role of intellectuals they cannot face the reality of their innermost craving. They need the sanction of a holy cause or an ideal before they can let themselves go. They will grab at any idea or word floating in the air and employ it as an incantation to conjure action out of the void. It is the presence of these activist intellectuals which renders words and ideas dangerous.

66

One might equate growing up with a mistrust of words. A mature person trusts his eyes more than his ears. Irrationality often manifests itself in upholding the word against the evidence of the eyes.

Children, savages, and true believers remember far less what they have seen than what they have heard.

67

Words have ruined more souls than any devil's agency. It is strange that the word, which is a chief ingredient of human uniqueness, should also be a chief instrument of dehumanization. The realm of magic is the realm of the invisible and the domain of the word.

68

It is a paradox of the post-industrial age that, despite its technical omnipotence, it is as dominated by words and magic as any primitive tribe. A haze of empty words, coming from the word factories of the universities, is corrupting the air of our ailing cities. The young lurch not so much from one illusion to another as from one cliché to another.

69

The oppressed and injured do have an advantage over the fortunate and the free. They need not grope for a purpose in life, nor eat their hearts out over wasted opportunities. Grievance and extravagant hope are meat and drink to their souls, and there is a hero's garment to fit any size, and an imperishable alibi to justify individual failure.

It is doubtful whether the oppressed ever fight for freedom. They fight for pride and for power—power to oppress others.

70

There has been a change in the tilt of the landscape of hope and faith. Formerly, dreams and ideologies flowed from the advanced to the backward countries. Now the flow is in the opposite direction. The faith and the hope afloat in the world at present come from backward countries. Such groups in the advanced countries as need faith, hope, and participation in a collective effort to make life meaningful must turn to the backward countries for inspiration.

71

The less we have, the more is there to be hoped for. Hence, where hope is an indispensable ingredient in bringing about social cohesion and discipline, a condition of perpetual scarcity might become part of governmental policy. In both Russia and China austerity might be a social necessity even when it becomes possible to produce a copious supply of consumer goods.

72

The contrast between the innate acquisitiveness of the Russian masses—their insatiable hunger for land—and the preaching of Communism is at the root of the present fanaticism in Soviet Russia. Any peasant society adopting Communism will display a fervent, intolerant state of mind for several generations.

73

It is probably true that business corrupts everything it touches. It corrupts politics, sports, literature, art, labor unions, and so on. But business also corrupts monolithic totalitarianism. Capitalism is at its liberating best in a noncapitalist environment. The crypto-businessman is the true revolutionary in a Communist country.

74

The revolutionary changes of our time are taking place in countries outside the Communist sphere. These countries have revolutions without revolutionaries. In the Communist world, where time stands still, there are self-styled revolutionaries but no revolutions.

It is indeed doubtful whether revolutionaries can

revolutionize their own country. The fateful effects of a revolution are usually felt elsewhere. The French Revolution altered France relatively little, but it created Germany. Similarly, the fateful effects of the Russian Revolution will be a United Europe and a new China.

75

It fares ill with the world when the strong imitate the weak, and the leaders of mighty Soviet Russia speak and act as if they were the leaders of "a company of poor men" surrounded by a hostile world.

The shifts and devices by which the weak turn weakness into strength become instruments of coercion and dehumanization when used by the strong.

III

Creators

76

Patience is a by-product of growth—we can bide our time when it is the time of our growth. There is no patience in acquisition or in the pursuit of power and fame. Nothing is so impatient as the pursuit of a substitute for growth.

77

They who lack talent expect things to happen without effort. They ascribe failure to a lack of inspiration or ability, or to misfortune, rather than to insufficient application. At the core of every true talent there is an awareness of the difficulties inherent in any achievement, and the confidence that by persistence and patience something worthwhile will be realized. Thus talent is a species of vigor.

78

People who cannot grow want to leap: they want short cuts to fame, fortune, and happiness.

79

No invention could ever take the hard work out of creating—out of good writing, painting, composing, inventing, etc. The economy of the spirit is incurably an economy of scarcity. An affluent society might be able to dispense with the ethic of work in its everyday life, but to attain any sort of excellence it will have to implant implacable taskmasters in the breasts of its people. Indeed, without the discipline of the creative effort the affluent society will be without stability. It might have to become a creative society in order to survive.

80

That which is unique and worthwhile in us makes itself felt only in flashes. If we do not know how to catch and savor the flashes we are without growth and without exhilaration.

81

A plant needs roots in order to grow. With man it is the other way around: only when he grows does he have roots and feels at home in the world.

82

Where thought is prompted by a penchant for weightiness and a high purpose the result is often a blend of pompousness and hysteria.

83

The compulsion to take ourselves seriously is in inverse proportion to our creative capacity. When the creative flow dries up, all we have left is our importance.

84

The discrepancy between trivial causes and momentous consequences is a crucial trait of human uniqueness, and it is particularly pronounced in the creative individual. It is a mark of the creator that he makes something out of nothing.

It has been said that all places are equidistant from heaven, and all eras equidistant from eternity. It is also true that all experiences are equidistant from a truth or regularity they illustrate. But it is to the creative mind only that a common occurrence can be as revealing as an outstanding event.

85

It seems that people who are occupied in doing what a society is particularly good at do not experience a marked decline in their creative powers with age. The writers of classical Greece kept producing great works in their eighties and nineties. In a military society generals remain young and brilliant to the end, just as in this country businessmen, technicians, and politicians are least subject to decline with age. The fact that in this country so many first-rank novel writers have petered out after forty suggests that America is not an optimal milieu for novelists.

86

It would be difficult to exaggerate the degree to which anonymous examples triggered creative outbursts or were the seed of new styles in the fields of action, thought, and imagination. A minor versifier, a minor composer, a mediocre writer, painter, or teacher, an untalented tinkerer may be the seed of momentous developments in art, literature, technology, science, or politics. Many who have shaped history are buried in unmarked and unvisited graves.

It is a mark of a creative milieu that lesser people can become instruments for things greater than themselves.

87

The genuine creator creates something that has a life of its own, something that can exist and function without him. This is true not only of the writer, artist, and scientist but of creators in other fields. The creative teacher is he who, in the words of Comenius, "teaches less and his students learn more." A creative organizer creates an organization that can function well without him. When a genuine leader has done his work, his followers will say, "We have done it ourselves," and feel that they can do great things without great leaders. With the noncreative it is the other way around: in whatever they do they arrange things so that they themselves become indispensable.

88

An empty head is not really empty; it is stuffed with rubbish. Hence the difficulty of forcing anything into an empty head.

89

What monstrosities would walk the streets were some people's faces as unfinished as their minds.

90

Familiarity blurs and flattens. Both the artist and the thinker are preoccupied with the birth of the ordinary and the discovery of the known. They both conserve life by recapturing the childhood of things.

91

When the genuinely creative imitate, they end up by making the model a poor imitation of themselves.

92

Language was invented to ask questions. Answers may be given by grunts and gestures, but questions must be spoken. Humanness came of age when man asked the first question. Social stagnation results not from a lack of answers but from the absence of the impulse to ask questions.

93

The Phoenicians invented the alphabet and the Greeks borrowed it from them. Yet how great the discrepancy between what the Phoenicians did with that which they originated and what the Greeks did with that which they borrowed. Perhaps our originality manifests itself most strikingly in what we do with that which we did not originate. To discover something wholly new can be a matter of chance, of idle tinkering, or even of the chronic dissatisfaction of the untalented.

94

The vigor of a society shows itself partly in the ability to borrow copiously without ill effects and without impairing its identity. The Occident borrowed profusely from other civilizations and thrived on it. It is startling to realize that between A.D. 1400 and 1800 the Eastern influence on the West was greater than the Western influence on the East. Had it not been for the Eastern influence, Columbus might not have set out to discover America. And it is well to remember that Asia gave us the instruments—gunpowder, the compass, the astrolabe—with which to subdue it.

It is a mark of debilitation that, at present, countries in Asia, Africa, and Latin America are sickening on their borrowings from advanced countries. Only Japan has been vigorous enough to borrow plentifully without getting social indigestion. Early in history, Egypt, Crete, India, and others borrowed freely from Sumer yet developed unique, vigorous civilizations.

95

Elitist intellectuals hug the conviction that talent and genius are rare exceptions. They are inhospitable to any suggestion that the mass of people are lumpy with unrealized potentialities. Yet there is evidence that the masses are a mine rich in all conceivable talents. We have as yet no expertise of talent mining but must wait for chance to wash nuggets out of hidden veins.

We know of a few instances where the masses were churned and scattered so that their talent content was made visible. The dumping of millions of common people from Europe on this continent was such a churning and scattering. A more sanguinary experiment on the masses was performed by Stalin in Russia. He liquidated the most cultivated segment of the Russian population and then proceeded to extract every conceivable talent and aptitude from the tortured, terrorized Russian masses. It was a cruel, wasteful way, yet no one will maintain that the Russian people are at present less endowed with talent than they were before the Revolution.

During the Renaissance there were instances of sudden activation of the creative powers of the masses. Most of the artists in Renaissance Florence were the sons of shopkeepers, artisans, peasants, or petty officials. The art honored in Florence was a trade, and the artists were treated as artisans.

Something similar happened in Holland during its golden century, 1600-1700. A nation of less than a million produced several thousand painters, among them Hals, Rembrandt, and Vermeer. After 1700 Dutch painting sharply declined and never recovered.

96

Elitists never tire of repeating that only the chosen few matter; the majority are pigs. Yet it does happen that a he pig marries a she pig and a Leonardo is born.

97

Right now there are no striking differences between the elites of different countries. One can visualize the writers, artists, scholars, scientists, and technologists of various nations assembled in one enclave, living together, not necessarily in perfect harmony but with no insurmountable barriers in ways of thought and of action. Obviously, the present great differences in vigor and welfare between nations cannot be the result of differences in the performance of their elites but must be ascribed to differences in the character of their masses.

98

To perform well, elites need tending and nurturing. They need attention, and would rather be persecuted than ignored. With the masses it is the other way around—like weeds they thrive best when left alone.

99

If, as seems to be true, vanity, malice, and envy are often accompaniments of creative striving, it should be possible to put up with them. The difficulty is that, although the truly creative are too few to form a class, we do have a marked-off segment of the population to which the writer, artist, scientist, etc., feel that they belong. This segment is made up of people who have much in common with the genuinely creative individual. They have the same sense of uniqueness and high purpose, and also the tendency toward malice, envy, backbiting, and self-dramatization.

100

If we want to delineate the typical state of mind of a group—its typical attitudes, aspirations, values, prejudices, biases—we must go not to the firmly established members of the group but to the least secure: those who, for one reason or another, feel themselves still as outsiders. It is those without an unquestioned sense of belonging who are most likely to exhibit all the essential identification marks, and know all the passwords.

101

Can there be a social order that will wholly suit the writer and the artist? There are so many contradictions between what they want and what is good for their creative flow. The writer and the artist need something to worship and something to resist. They need praise and rewards, but they also need to be left alone to stew in their own juice.

102

The sense of uniqueness inherent in the creative process has often inclined the writer and the artist to see themselves as the center of the universe and as the bearers of a destiny shaped by cosmic forces. Hence their fascination with coincidences, omens, and signs. It is a conceit which requires a high capacity for self-dramatization—a capacity indigenous to the juvenile mentality. It is amazing how much phoniness is needed to produce a grain of originality.

103

Both the revolutionary and the creative individual are perpetual juveniles. The revolutionary does not grow up because he cannot grow, while the creative individual cannot grow up because he keeps growing.

104

The contemporary explosion of avant-garde innovation in literature, art, and music is wholly unprecedented. The nearest thing that comes to mind is the outburst of sectarian innovation at the time of the Reformation, when every yokel felt competent to start a new religion. Obviously, what our age has in common with the age of the Reformation is the fallout of disintegrating values. What needs explaining is the presence of a receptive audience. More significant than the fact that poets write abstrusely, painters paint abstractly, and composers compose unintelligible music is that people should admire what they cannot understand; indeed, admire that which has no meaning or principle.

105

Total innovation is the refuge of the untalented and the innately clumsy. It offers them a situation where their ineptness is acceptable and natural. For we are all apprentices when we tackle the wholly new, and we expect the new to show the apprentice's hand—to be clumsy and ill shapen.

Yet, however untalented and clumsy, the innovators have a vital role to play. For it is the fate of every great achievement to be pounced upon by pedants and imitators who drain it of life and turn it into an orthodoxy which stifles all stirrings of originality. The avant-garde counteracts this deadening influence, and fulfills the vital role of keeping the gates open for the real talents who will eventually sweep away the inanities of the experimenters, and build the new with a sure hand.

106

A true talent will make do with any technique.

107

How much easier is self-sacrifice than self-realization!

108

It is the stretched soul that makes music, and souls are stretched by the pull of opposites—opposite bents, tastes, yearnings, loyalties. Where there is no polarity—where energies flow smoothly in one direction—there will be much doing but no music.

109

The real antichrist is he who turns the wine of an original idea into the water of mediocrity.

IV

Prognosticators

110

There has been a gradual narrowing of the range of predictability during the past five hundred years. In the heyday of Christianity, predictability reached the utmost limit—the life beyond. In the idea of progress, which took the place of millennial prognostication, the range of predictability was narrowed to a century or so. With the end of the First World War, predictability shrank further: the craving for security took the place of hope, and people were satisfied if they could foresee the course of a single lifetime. If the shrinking continues, we shall be satisfied if we can predict in the evening the eventualities of next morning. This has already happened in some totalitarian countries, where a man considers himself fortunate if he can be certain that he will not be imprisoned, exiled, or liquidated between going to bed and getting up.

In the past, societies with a vivid conception of a life beyond were indifferent to divination and prophecy. The ancient Egyptians, who expended much treasure and effort in preparing for a hereafter, did not develop any sort of astrology, while the Babylonians, who had no faith in a life beyond, cultivated divination. Hebrew prophecy was at its height when resurrection was not as yet an article of faith. In Europe, astrology came into prominence during the Renaissance when millennial Christianity was losing its hold on the educated.

It is a paradox that in our time of rapid, drastic change, when the future is in our midst, devouring the present before our eyes, we have never been less certain about what is ahead of us. Our need for predictability is far more urgent than in times past, and we are addicted to forecasters and pollsters. Even when the forecasts are wrong we go on asking for them. We watch our experts read their graphs the way the ancients watched their soothsayers read the entrails of a chicken.

111

Absolute power can make people predictable; it can turn human variables into constants. Engineers of souls like Stalin and Hitler can make of history an exact science—a branch of zoology.

112

When a Stalin or a Hitler can predict the future because he has the power to make his predictions come true, the life of the average man becomes unpredictable. It is with prediction as with wealth: there is so much of it in a society, and when one person has most of it there is little left for others.

113

It is remarkable how little history can teach us at present. The past seems too remote and different to matter. We can obtain insights about the present not from books of history but from books dealing with the human condition. It is becoming evident that the more technology triumphs, the less do things and impersonal factors shape human events. The post-industrial age will be dominated by psychological factors, and a meaningful history of our time must base itself on the assumption that man makes history.

114

Were history mainly a record of accidental and deliberate experiments performed on mankind, it would be to an understanding of human nature what laboratory experiments are to an understanding of nature. Such a history could be an aid to an understanding of ourselves.

115

History does not repeat itself. There is no more reason why one era should recapitulate the events of another era than that the literature or the art of one period should be duplicated in another. But the conditions which favor the birth of literature and art do repeat themselves, and so do the factors which promote social vigor, stagnation, stability, ferment, and so on.

116

We know now that when dreams come true they may turn into nightmares. And the paradox is that now that we know the end of the story we are totally in the dark about the future. The writing on the wall is there for all to see, but no one has found the key to decipher it.

117

Every passionate search is in some degree a search for something lost. Even the search for the wholly new often starts out as a search for a substitute for something lost. The voyages of discovery and exploration which marked the birth of the modern Occident were to some extent a corollary of a loss of faith in a heavenly kingdom. The explorers who searched for new continents, fabulous empires, and magic islands were constantly on the lookout for signs of paradise—a heaven on earth.

118

The inability to know what is happening in the present is an aspect of a wider phenomenon: we know very little about ourselves—what we look like, how we sound, what is really going on inside us. We need insights about the things that are happening under our noses; we have to divine them, and we are most credulous about them.

119

It could be that human nature is stubbornly resistant to drastic change. Hence the fact that they who set their hearts on realizing revolutionary changes are as a rule hostile to human nature; they become antihuman, so to speak. They will do all they can to turn men into soulless material.

120

The birth of the new constitutes a crisis, and its mastery calls for a crude and simple cast of mind—the mind of a fighter—in which the virtues of tribal cohesion and fierceness and infantile credulity and malleability are paramount. Thus every new beginning recapitulates in some degree man's first beginning.

121

We have perhaps a natural fear of ends. We would rather be always on the way than arrive. Given the means, we hang on to them and often forget the ends.

122

The unpredictability inherent in human affairs is due largely to the fact that the by-products of a human process are more fateful than the product.

123

In our time one has to ask children and the ignorant for news of the future. The saying in the

Talmud that after the destruction of the temple prophecy was taken from the wise and given to children and fools reflects the disarray and perplexity of a time of trouble. When things become unhinged, wisdom and experience are a handicap in discerning the shape of things to come.

124

The ignorant are a reservoir of daring. It almost seems that those who have yet to discover the known are particularly equipped for dealing with the unknown. The unlearned have often rushed in where the learned feared to tread, and it is the credulous who are tempted to attempt the impossible. They know not whither they are going, and give chance a chance. Often in the past the wise were unaware of the great mutations which were unfolding before their eyes. How many of the learned knew in the early decades of the nineteenth century that they had an industrial revolution on their hands? The discovery of America hardly touched the learned but inflamed the minds of common folk.

However much we talk of the inexorable laws governing the life of individuals and of societies, we remain at bottom convinced that in human affairs everything is more or less fortuitous. We do not even believe in the inevitability of our own death. Hence the difficulty of deciphering the present, of detecting the seeds of things to come as they germinate before our eyes. We are not attuned to seeing the inevitable.

V

The Individual

126

It is doubtful whether there can be such a thing as collective shame. Collective resentment yes, and of course collective pride and elation—but not shame. The association with others is almost always felt as an association with our betters, and to sin with our betters cannot be productive of a crushing feeling.

127

It is not sheer malice that pricks our ears to evil reports about our fellow men. For there are frequent moments when we feel lower than the lowest of mankind, and this opinion of ourselves isolates us. Hence the rumor that all flesh is base comes almost as a message of hope. It breaks down the wall that has kept us apart, and we feel one with humanity.

128

Retribution often means that we eventually do to ourselves what we have done unto others.

129

There is probably an element of malice in the readiness to overestimate people: we are laying up for ourselves the pleasure of later cutting them down to size.

130

How frighteningly few are the persons whose death would spoil our appetite and make the world seem empty.

131

We tend to exaggerate not only the sins of others but also their remorse, sensitivity, gratitude, attachment, hatred, etc. In short, we usually see the peculiarities of others as through a magnifying glass; we also see ourselves in exaggerated proportions when we see ourselves through the eyes of others. We attach a quality of extremism to the opinion other people have of us.

132

Sometimes it seems that people hear best what we do not say.

133

The capacity for identifying ourselves with others seems boundless. No matter how meagerly endowed, we yet find it easy to identify ourselves with persons of exceptional endowments and achievements. Can it be that even in the least of us there are crumbs of all abilities and potentialities so that we can comprehend greatness as if it were part of us?

134

We can be vividly impressed only by what we are attuned to—by anything in the outside world which has some counterpart inside us. Hence it is that the old, so much more than the young, are poignantly aware of the inexorable wear and tear that is going on in the world around us.

135

We find it hard to apply the knowledge of ourselves to our judgment of others. The fact that we are never of one kind, that we never love without reservations and never hate with all our being cannot prevent us from seeing others as wholly black and white.

136

A feeling of utter worthlessness levels a man's attitude toward his fellow beings. He views the whole of humanity as being of one kind. He will despise equally those who love him and those who hate him, those who are noble and those who are mean, those who are compassionate and those who are cruel. It is as if the feeling of worthlessness cuts one off from the rest of mankind. One sees humanity as a foreign species.

137

How often are we offended by not being offered something we do not really want!

138

Sensuality reconciles us with the human race. The misanthropy of the old is due in large part to the fading of the magic glow of desire.

139

Some persons must love and be loved in order to be intelligent.

140

What are we when we are alone? Some, when they are alone, cease to exist.

141

People who bite the hand that feeds them usually lick the boot that kicks them.

142

There are no chaste minds. Minds copulate wherever they meet.

143

How many and deep are the divisions between human beings! Not only are there divisions between races, nations, classes, and religions but also an almost total incomprehension between the sexes, the old and the young, the sick and the healthy. There would be no society if living together depended upon understanding each other.

144

No matter what our achievements might be, we think well of ourselves only in rare moments. We need people to bear witness against our inner judge, who keeps book on our shortcomings and transgressions. We need people to convince us that we are not as bad as we think we are.

145

What we are looking for is not people who agree with us but people who think well of us and know how to express it. We cherish such people though they disagree with us.

146

When people do us good our exhilaration is due not merely to the good we receive. In addition we feel that we are on the right path, that we have chosen well to be where we are. We see the good that happens to us as a good omen.

147

It is a sign of a creeping inner death when we no longer can praise the living.

148

When we begin to think that most people are no better than we, the world seems full of people who are fairly unpleasant.

149

The fundamental indifference toward us of most people we meet is a fact we are extremely reluctant to accept. Yet it should serve as a basis for our fair dealing with others: we must expect little, and we must give people some sort of vested interest in us to rouse and hold their interest.

150

The end comes when we no longer talk with ourselves. It is the end of genuine thinking and the beginning of the final loneliness. The remarkable thing is that the cessation of the inner dialogue marks also the end of our concern with the world around us. It is as if we note the world and think about it only when we have to report to ourselves.

151

So true is it that the path of desire once trodden remains frequented that we not only keep wanting what we cannot have but go on wanting what we no longer really want.

152

Were the world to treat us the way we treat ourselves we would turn into firebrand revolutionaries.

153

We need not only a purpose in life to give meaning to our existence but also something to give meaning to our suffering. We need as much something to suffer for as something to live for.

154

A man's heart is a grave long before he is buried. Youth dies, and beauty, and hope, and desire. A grave is buried within a grave when a man is buried.

155

So long as our capacity to savor a fulfillment is unimpaired, we keep on trying no matter how numerous the misses—we cannot learn from experience. It is only when a fulfillment no longer brings a singular joy that the slightest disappointment can teach us a lesson for good.

156

The feeling of being hurried is not usually the result of living a full life and having no time. It is on the contrary born of a vague fear that we are wasting our life. When we do not do the one thing we ought to do, we have no time for anything else —we are the busiest people in the world.

157

Our achievements speak for themselves. What we have to keep track of are our failures, discouragements, and doubts. We tend to forget the past difficulties, the many false starts, and the painful groping. We see our past achievements as the end result of a clean forward thrust, and our present difficulties as signs of decline and decay.

158

How easy it is for a failure to seem foolish!

159

It is loneliness that makes the loudest noise. This is as true of men as of dogs.

160

Despite our self-righteousness, we feel the good that happens to us as undeserved.

161

It takes a leaden weight off our backs to remember how unworthy we are.

162

We are more prone to generalize the bad than the good. We assume that the bad is more potent and contagious.

163

To the excessively fearful the chief characteristic of power is its arbitrariness. Man had to gain enormously in confidence before he could conceive an all-powerful God who obeys his own laws.

164

A sensitive conscience is often a by-product of a decline in vigor. When we are growing, our doings are transitory, mere stepping stones to be left behind, but when we stop growing we are what we do and think.

165

To grow old is to grow common. Old age equalizes—we are aware that what is happening to us has happened to untold numbers from the beginning of time. When we are young we act as if we were the first young people in the world.

166

It needs some intelligence to be truly selfish. The unintelligent can only be self-righteous.

167

Sometimes we feel the loss of a prejudice as a loss of vigor.

168

The indisputable fact that we do not, and perhaps cannot, recognize our own voice indicates how incurably strange we are to ourselves.

169

We are more surprised when something we expected comes to pass than when we stumble on the unexpected.

170

It sometimes seems that the thing we least possess and can call our own is our self. We cannot be sure of our faculties, talents, and creative powers. We can possess and keep under lock and key only that which is not part of the self.

171

To have an exceptional talent and the capacity to realize it is like having a powerful appetite and the capacity to enjoy it. In both cases there is an impatience with anything that hampers free movement, and the feeling that the world is one's oyster.

172

The hardest arithmetic to master is that which enables us to count our blessings.

173

We never say so much as when we do not quite know what we want to say. We need few words when we have something to say, but all the words in all the dictionaries will not suffice when we have nothing to say and want desperately to say it.

174

The remarkable thing is that it is the crowded life that is most easily remembered. A life full of turns, achievements, disappointments, surprises, and crises is a life full of landmarks. The empty life has even its few details blurred, and cannot be remembered with certainty.

175

A man's worth is what he is divided by what he thinks he is.

176

Our triviality is proportionate to the seriousness with which we take ourselves.

177

Doing nothing is harmless, but being busy doing nothing is not.

178

Our greatest weariness comes from work not done.

179

We can remember minutely and precisely only the things which never really happened to us.

180

Much of the time we hate most truly people we have never met.

181

It is surprising how much truth there can be in the most malicious guesses people make about us.

182

Often, that which strikes us as the defeat of a hope is actually its fulfillment.

183

It is the individual only who is timeless. Societies, cultures, and civilizations—past and present—are often incomprehensible to outsiders, but the individual's hungers, anxieties, dreams, and preoccupations have remained unchanged through the millennia. Thus, we are up against the paradox that the individual who is more complex, unpredictable, and mysterious than any communal entity is the one nearest to our understanding; so near that even the interval of millennia cannot weaken our feeling of kinship. If in some manner the voice of an individual reaches us from the remotest distance of time, it is a timeless voice speaking about ourselves.